Physics for Everyone

Quick Basics

Physics can be fun.
Trust me!

Written By Vedang Sati

CONTENTS

What is Physics?

We realize, at some point in our lives, that nature is full of patterns and to read these special arrangements we use our keen observation and a distinguished language, the language of mathematics. But what is physics? Is it a complex analysis relying entirely on mathematics? Or is it a purely observational science?

Without observation, there is no chance of analysis! So we can say that physics is a combination of both. Even Isaac Newton had to observe the fall of an apple in order to make any progress! Observation includes asking questions with an open mind. So Mr. Newton asked, *"Why did the apple fall down?"* – A very simple question which other people might have considered stupid: *"Isn't it obvious?"* and *"What's so special about it?"* – They would have said…

But that simple question alone, made one ordinary farmer perhaps the greatest, and the most respected scientist in human history. It is only by asking questions that we have made progress. In physics we ask questions that start with *what, how and why*. Then we make deductions to get closer to the truth.

Every natural phenomenon hides some sort of secret. We try to find these pieces of a large puzzle – *our universe* – and join these pieces one by one to get the big picture, which is our goal. This big picture, we believe, will be governed entirely by the laws of physics. Will we ever be able to accomplish this goal? We don't know that yet! But we have to keep learning from nature. Who knows what we may find in the near future...

Physics is not just a subject to study, it is not just about grades and passing the course – physics is about how our universe works...

Why you should read this book

Physics is the study of *almost* everything – from the very small *(atom)* to the very large *(universe)!* But in order to understand the complicated things, we need to learn the really basic things first.

This book is a quick read. Very interesting concepts, explained with the help of illustrations, activities and fun. We believe that physics is beyond equations, it is in our everyday experience. It is hard for most people to enjoy physics. But **we** do things differently. We want you to have fun while you are learning because *what we learn with pleasure, we never forget…*

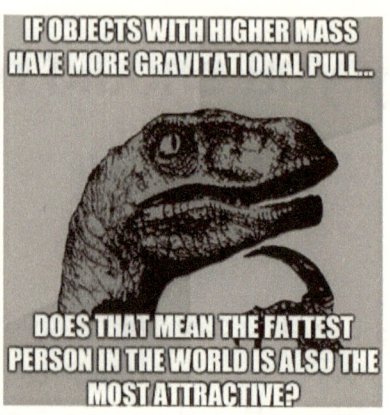

IF OBJECTS WITH HIGHER MASS HAVE MORE GRAVITATIONAL PULL...

DOES THAT MEAN THE FATTEST PERSON IN THE WORLD IS ALSO THE MOST ATTRACTIVE?

In this short book, we will learn the most basic physics concepts! And we are certain that you will enjoy the most while you tackle the world of physics. We will be looking at physics in our everyday lives and real-life applications (*motion, gravity, energy, light* and *atoms*). We make learning easy and entertaining at the same time. Read this book and *fall in love* with physics...

...One more time!

Understanding Motion

Certainly all objects are moving!

That book on your table may seem stationary *to you* but since our planet Earth is moving, the book is also moving! It is all a matter of perspective. Someone sitting on the moon definitely does not agree with your perspective!

Even atoms are in continuous motion! They are jiggling, vibrating, splitting and what not! Everything is moving. But why do things move in the first place?

Motion is a part of our everyday lives. All objects that move have speed. Some are very fast *(cheetah)* and others are slow *(turtles)*. By knowing how fast something is moving, we are able to tell what its speed is.

Our ancestors have always tried to understand motion. Centuries ago, they made efforts to understand macroscopic motion *(motion of things that we see with naked eyes)* and these days we are trying to grasp the motion of microscopic particles *(like atoms)*.

Let's go back in time.
Aristotle, a Greek philosopher, was once asked,

"Why do things move?"

He replied,

"When you apply force on objects, they move!"

Clearly, when you kick a ball, it moves! And so people believed Aristotle. Aristotle further said that all the heavenly bodies like the moon and the sun (*and others*) moved around the majestic planet of Earth!

Clearly you see the sun and the moon cross our skies every day and every night. So far so good. People still believe in Aristotle's views.

People then asked,

"Why do the moving things stop?"

And Aristotle responded,

"Isn't it obvious? Things stop moving when they get tired. Don't you stop moving when you are out of energy?"

Clearly Aristotle was right and people became Aristotle's ardent supporters. His fan base kept on growing!

But much later, one Italian man dared to challenge the authority of Aristotle. His name was Galileo. Galileo wanted to test each and every statement that Aristotle ever said…

Galileo observed that it was hard for anyone to move a heavy rock. But anyone could lift a book. He introduced a new property associated with an object's mass, *Inertia*.

Galileo said that heavy bodies had more inertia and lighter objects had less inertia. He also said,

*To move a body with high
amount of inertia is more
difficult than disturbing a
body with less inertia.*

Clearly it is more difficult to lift an elephant than lifting a puppy.

Now we want to introduce analogy. Compare any object's inertia with laziness. Assume that all the objects are lazy and the heavier they are, the lazier they become! Simple!

So according to Galileo, all objects resist change in their current state *because they are lazy.* Their resistance to change depends on the amount of inertia, which in turn, depends on the amount of mass they possess.

So if an object is moving, it will not want to stop and if *(for god's sake)* we wish to stop it, we will need to **fight** its inertia!

Hmm...That is why it is easier to stop a moving bicycle than stopping a giant-fully-loaded truck! We experience inertia everyday – when a car is moving, all its passengers are also moving along with the car. However, when the car suddenly stops, we fall forward.

This is because of inertia – our bodies wanted to keep on moving, even when the car had stopped. That is why, there are seat-belts that keep us sticking to the seats to avoid the chance of an accident. Moreover, there are air bags placed for the driver so that he doesn't bang his head on the dashboard!

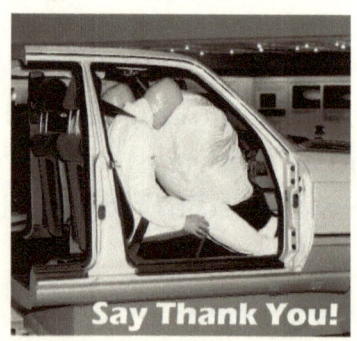

Say Thank You!

So, a moving object does not want to stop and, a stationary object does not want to move. Who is responsible for this behavior?

Blame inertia!

One day, Galileo observed a leaf fall off a tree. He noticed its crooked trajectory, and compared it with a falling stone. He realized something very important!

Galileo understood that 'something' was stopping the leaf from falling. On the other hand, he saw that the stone followed a straight line as it fell from certain height. He quickly recognized that *air* was offering resistance!

He then asked himself a new question,

> *"What if, there is resistance offered by the floor when things are moving!?"*

He understood that objects did not stop because they were tired, they stopped because something was causing trouble when they moved. He called that troublemaker *friction.* He started to design smooth surfaces to test his argument.

Friction is least when objects are spherical and rolling. So he used metallic spheres and rolled them down an inclined plane.

He observed that as soon as the sphere reached the bottom of the incline, it travelled in a straight line and eventually stopped. He conducted another experiment in which he used two inclined planes facing each other (a) and (b), as shown below:

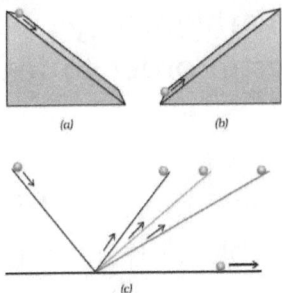

He started rolling the marble from some height 'h'. The marble reached the bottom, continued its journey and climbed up the same height 'h' on the second inclined plane – as shown in (b)

As Galileo reduced the angle of inclination of the second inclined plane, the marble travelled further in order to reach the same height 'h' – as shown in (c). He asked,

"What would happen if he decreased the inclination angle of the second inclined plane to zero?"

Clearly the marble will get 'confused' and travel forever in order to reach the same height 'h'.

But wait! Since there is friction on the floor, the marble will eventually come to a halt!

Galileo also invented the telescope and became the first man to observe the universe! He saw 4 small objects encircling the planet of Jupiter. He realized that Aristotle was wrong.
Again!
Aristotle had assumed that all objects moved around planet Earth, but Galileo had discovered that it was otherwise.

Some men just want to watch the world turn...

Galileo paid a heavy price for opposing Aristotle, but that is another interesting story...

Now we move ahead in time and meet an English scientist, Newton! Sir Isaac Newton built on Galileo's results and presented very important laws of **motion.**

The **first law of motion** is the law of inertia. Objects are lazy. They do not want change in their state unless and until someone from the outside brings that change. This external agency bringing the change is called *force*.

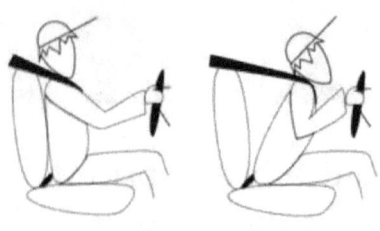

Newton's **second law of motion** tells us what happens when we apply force on a body. It says that *velocity* of the body will change. This means that either the object will change speed, or it will undergo a change in direction.

When we apply brakes to stop a moving car, we are actually slowing it down. Thus we are applying force on the car to make it stop! When a footballer kicks a ball, he not only changes the speed of the ball but also gives it a new direction. This is an example when force changes the direction of motion.

Then there is the **third law of motion** which says that for every action there is an equal and opposite reaction. If body A applies force F on body B, then *automatically* and *instantaneously* body B applies the force F on body A in the opposite direction. Jeez, this is getting confusing! Let us try something else…

First of all, we need to understand that action and reaction occur together. It is an instantaneous process. Second, we also need to know that action and reaction occur on two different bodies. Third thing is that action and reaction are equal in strength but opposite in direction. Now consider the image below:

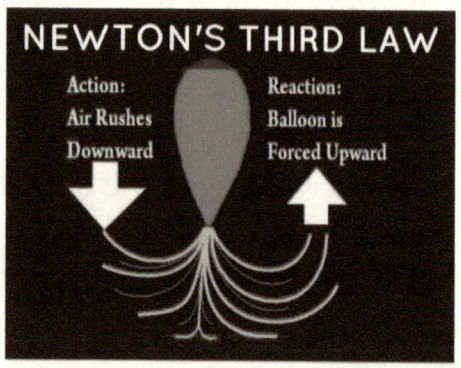

This makes it simple to understand!

We still do not understand motion completely! Yes we know the 3 laws of motion, but they are certainly not enough! There is one more interesting thing left: *relativity*.

The book on your table is still there and not moving. Or is it!? Yes this has tricked Aristotle before. He saw the moon and the sun moving on a daily basis, but there was not any way he could see the 'behind-the-scene' motion of his own planet!

Who is to blame?
Relativity!

Yes the book is appearing stationary to you. But for some astronaut on the moon, you along with the book, are moving at a tremendous speed around the sun! So it is all a matter of perspective. This difference is called *relative motion.*

Two cars are moving side by side at the same speed in the same direction. Try this yourself! Passengers in car A observe passengers in car B. And they notice that they are not moving with respect to one another!

Believe me. It works!

Story of Gravity

Tons of ocean water is held together by an enormous force of gravity. You also stick to Earth because of gravity. When you kick a ball, you apply force by touching the ball – this is called *contact force*. Gravity, on the other hand, is an invisible force between masses.

We all know the story of gravity. One fine day, Isaac Newton was out in the open when suddenly, he saw an apple fall off a tree! He asked a very simple question,

"Why did the apple fall down?"

But the story of gravity is much older than Isaac Newton. And before we understand what Newton did, we need to understand what others did to support his work…

The story starts with Aristotle. He said that all bodies moved around Earth. Why did they move around earth? He had no answer. But people were die-hard fans. They believed in Aristotle's words.

Then came a Polish priest and astronomer, Copernicus. He said that earth moved around the sun in a circle. He published his ideas anonymously for he was afraid to do it openly – he was a religious priest! But people found the truth eventually. He was locked up in the prison.

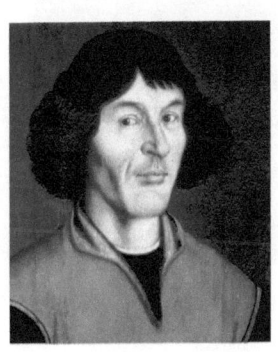

Copernicus did not have a solid proof, maybe that's why he was afraid. But he found one sincere supporter in Johannes Kepler, a German mathematician and astronomer.

Kepler deduced important laws of planetary motion. He said that planets moved around the sun in elliptical orbits. An ellipse is an elongated circle.

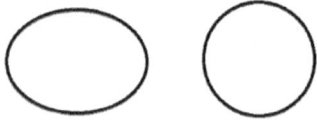

Although Kepler used astronomical data and geometry to develop his theory, he did not present any solid proof of his own.

Decades later, Galileo invented the telescope with which he could see the planets in motion. But he also saw that smaller bodies were moving around Jupiter. He already knew that the *moon* revolved around Earth. He realized that there was *some sort of connection* between the two observations.

What was it?

17th century, a fine sunny day, young Isaac Newton is enjoying the sunshine, when suddenly, he sees an apple fall down. He is confused.

Why didn't the apple go up?

What pulls the apple down?

Clearly someone is pulling the apple down.

It has to be earth!

But there was a problem…

Newton had no idea how to represent his discovery on paper! He needed equations to determine the motion of any falling apple. Observing nature is not just enough, he had to convert his idea into something meaningful. Because why would other scientists and mathematicians believe on a poor farmer?

He spent years to create *a new branch* of mathematics now known as calculus. Many people say that it was Leibniz (a German mathematician) who first created calculus. Other say Newton and Leibniz both did it at exactly the same time.

Whatever!

Newton used mathematics to understand the motion of a falling body. He explained motion, and most importantly planetary motion.

Newton used the results of Copernicus, Kepler and Galileo to formulate new ideas. He said,

> *"If I have seen any further, it is by standing on the shoulders of giants!"*

The universal law of gravitation states that *gravity* is a fundamental and universal force, i.e. all the bodies in the universe feel this force.

The force of gravity gets stronger between massive bodies, and weaker between small lighter objects. Like electrons and atoms do not even feel this force! They are incredibly light! On the other hand, planets and stars are dominated by this force alone.

$$F \propto \frac{m_1 m_2}{r^2}$$

In fact, we stick to Earth because of gravity! Enough said, this is the story of gravity. But wait, another important update came along...

Albert Einstein recognized that gravity was not *just a force!* In fact gravity could be thought of as waves! But this concept is a little too complicated. And this book is only *basic*...

Everything is Energy!

And that is all there is to it. What is energy? You eat food, you get the power to do work. If you do not eat food for a few days, you become weak and pale. Food is actually stored *chemical energy*. Plants use *light energy* and store this energy in small *blobs* that we call *fruits* or *vegetables*.

Actually the entire universe is governed by exactly the same principle. Energy in one form is undergoing a change to reappear in second form. But it is still there! It does not vanish! And no new energy appears! Everything remains the same.

But still, we did not answer what energy actually is! What is it?

Suppose there is a box. An empty box. A very unattractive box. Nobody gives a damn whether it exists or not. But one day, an alien comes along and fills the box with chocolates, lots of them!

Now there are chocolates in the box, and people who earlier paid no attention whatsoever, are suddenly attracted to this box. Yes! Because now there are chocolates & people are bound to get closer. There was a time when this same box was empty, unhappy and powerless. And now the box has gained this magical ability to *"move"* people!

Similarly, energy has the ability to move all kinds of objects. Energy is like chocolate! Kinetic energy gives objects speed. Sunlight or solar energy makes plants happy. Food energy attracts people. Heat energy moves particles, they literally start dancing/vibrating!

So this is energy.

In fact, you are an incredible amount of energy! *Mass* is also energy. Even a tiny atom is source of a large amount of energy. Ever heard of atomic bombs?

Yes, energy can be dangerous too, if not handled properly...

Conversion of energy is very useful too. Like plants convert sunlight into food. Solar panels convert sunlight into electricity. A tungsten light bulb converts heat into light. A dam converts potential energy into electricity.

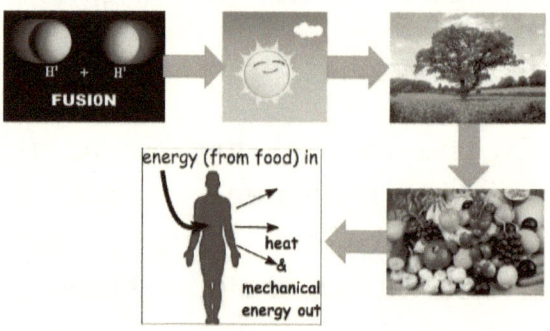

Let us talk about different forms of energy…

Heat is a form of energy. When you enjoy a sunny day, you are actually feeling sun's heat that has travelled millions of miles of distance. When you leave a hot cup of tea unattended for some time, it gets cool. Your tongue stays safe.

These are all processes. Some bodies are hot and some are cold. But that is all because of perspective – a hot cup of tea is cool when compared with the surface of a typical star. Also, *cold* is actually absence of heat!

But why do objects get hot? And why do they have heat energy? Actually particles are continuously in motion. They are jiggling and vibrating without getting tired. Cold ice cubes are very calm. But boiling water is not! You have seen this with your own eyes!

When an object is heated up, particles start to jiggle more and more! They start vibrating/dancing. On the other hand, when you cool them they get less excited and there is less dancing/vibrating.

Particles on the surface of a hot beverage are the most excited ones. They are very energetic. They jiggle so much that if you blow air they fly away! Literally they fly away, taking with them, all the energy! And so, your hot beverage gets less hot when you blow it with your mouth!

Now let us talk about **potential energy.** As its name suggests, something has *the potential* to do some task, but is not currently doing it! That hot cup of tea is kept on a table. It is safe for everyone. What if someone takes this hot cup of tea to a certain height? Then there is potential energy. He/she might let go of the cup and spill the hot beverage on someone's head!

Unlucky!

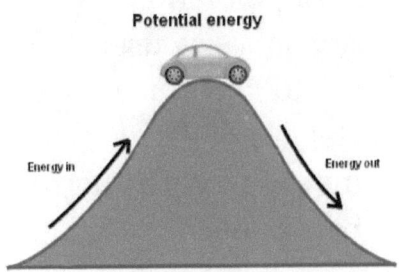

Potential energy

Energy in

Energy out

Yes, but when the cup is falling down, it gains speed right? This is **kinetic** energy. Moving objects have speed by virtue of which they have kinetic energy. When an object falls from a certain height its potential energy changes into kinetic energy. And when it gets to the ground, the potential energy becomes zero. And the ground gets *hurt*.

More the speed, more the kinetic energy. Very good!

Conversion of potential energy into kinetic energy is very useful. Dams use this method to produce hydroelectric power! Water is falling from certain height (dams are very high) and its energy is being used to make electric power!

This is the future of energy!

Okay! Great!

You may have played a lot with springs. What do you do? You stretch or compress the spring, and *let go!* The spring starts to go crazy! This is because when you compressed or stretched the spring, it got the hold of **elastic energy.** Elastic energy gets *stored.* A very special power!

The spring remembers that it was compressed or stretched, it then behaves accordingly.

There are other forms of energy like electrical energy, light energy, nuclear energy. But they are a little more advanced concepts...

Welcome to the atomic world!

Paper looks smooth, right? Clothes look smooth too, right? So on the outside everything looks smooth. But what will things look like when we see them under a microscope? We will first encounter microscopic world – the place where microbes live. If our microscope is powerful enough, we will magnify into the atomic world – the place where electrons and other particles live in harmony! Before discovering electrons and other particles, we believed that atoms were indivisible...

Nobody knows what an atom looks like. But we curious people try to imagine a few things about atoms. Like they are spheres. Like they may also have an internal structure!

And for decades we have tried to understand the mysterious atomic world.

Democritus first suggested that materials were not smooth but in fact were made up of *indivisible* atoms. So he suggested that atom did not have any internal structure. He also said that different objects were made of different types of atoms.

So far, so good.
But Joseph Thomson, an English scientist discovered a mysterious particle inside an atom whose mass was much less than the entire atom! He called it *electron*.

Electron is a charged particle. Just like particles have mass, particles *may* also possess charge. But unlike mass, charges are fixed numbers.

After discovering electron, a sub-atomic particle, Thomson had broken the age-old belief that atoms did not have internal organs.

Electron was the first organ discovered!

Now Thomson wanted to design this atomic body, just like biologists design human body. The only difference is that physicists use only *wild* imagination. Unlike biologists, physicists do not have tools to look at the inside structure of an atom.

Since bodies are neutral as a whole, he realized that there must be a positively charged particle (proton) inside the atom. Positive and negative would cancel out and make things neutral!

Plum Pudding Model

This is Thomson's model of atom. You can see negative charged electrons embedded within a positive sphere. The number of positives and negatives are exactly the same so that the thing remains **neutral.**

But one day, Rutherford wanted to test this model. Yes! In physics scientists are continuously challenging other scientists! Rutherford was a brilliant scientist. He, along with his students, designed a very simple experiment. They decided to shoot alpha ray at an atom (gold)! Alpha ray is a beam of alpha particles and alpha particle is just a positively charged particle much bigger than a proton.

They had to observe how alpha beam interacted with gold atoms. For that they used a detecting screen.

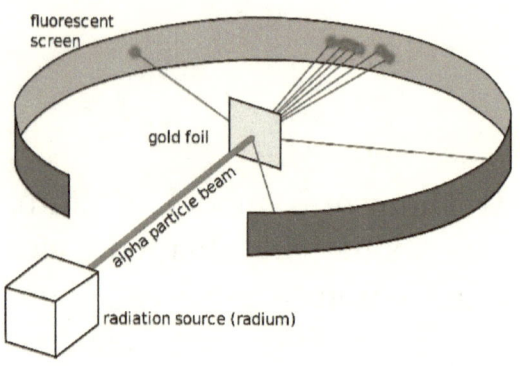

When they compiled their results they found:

1. Most of the alpha particles passed straight through the gold atoms
2. A few alpha particles were deflected

The first result implied that most of the atom was empty. Thomson had suggested a rather full and dense structure. But Rutherford's experiment acted otherwise – if a ball hits a dense and solid wall, it rebounds every single time. They found that almost all the alpha particles passed straight through the gold foil!

Thus gold atoms must be empty.

The second result of the experiment implied that a few places inside the atom were occupied by electrons. Because positive and negative attract, some alpha particles would have felt attraction and deviated from their paths.

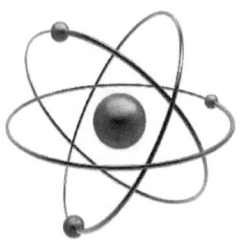

This is Rutherford's model of atom in which there is mass at the center, and electrons revolve around that mass. And all of the space between them is completely empty. Rutherford is so happy! But another physicist Niels Bohr found a severe error in Rutherford's model. Let us talk about that in the next chapter...

Electromagnetism

Take a balloon and rub it with your dry hair for some time! Wait a moment, you impatient child!

Now it will stick on the wall. There is no glue required. This is happening because the balloon is now *charged up*. This is called charging by rubbing. This is static electricity.

Do you have a cat?

In the above image, playful cat is stuck to a balloon due to static electricity. You know this, right!?

You may have played with magnets too! They are small objects that stick with each other, and they attract a few metals too! This is called magnetism.

For centuries, electricity and magnetism were thought of as different phenomena. But it was realized that they were in fact two aspects of the same phenomenon. We call this phenomenon *electromagnetism*.

Hans Oersted was conducting an experiment with electricity. A magnetic compass was kept in close proximity. Whenever current passed through the wire, the magnetic needle started going crazy!

What was happening?

Oersted concluded that the electric current was interacting with the magnetic field of compass. But since only a magnet can interact with another magnet, he supposed that the current was producing a wave of magnetism in its vicinity...

Story of electromagnetism begins.

Hans Oersted discovered that electricity could produce magnetism. Another man called Michael Faraday discovered that magnetism could in turn re-produce electricity!

Thus electricity and magnetism were actually the two sides of the same coin! It was James Maxwell who correctly compiled all the results in the language of mathematics. He also found that light was an electromagnetic wave. What we see as red or green or blue is nothing but electricity and magnetism!

Interesting!

This remained a theory until,

Heinrich Hertz, a German experimental physicist, finally created electromagnetic waves (radio) for real, by an accelerating charged particle.

Now let's get back to Rutherford's model of atom which said that electrons moved around a dense mass (nucleus) in circles. Electrons are charged particles. And whenever a charged particle is under acceleration, electromagnetic waves were produced as Hertz had already proved.

Going in a circular orbit around the nucleus is a difficult task. At each and every point of the circular trajectory, the particle has to change its direction. Since there is a change in direction, there is a continuous change in velocity.

Thus electron, the charged particle, is accelerating around the nucleus! Therefore it must produce electromagnetic waves as it revolves. This was first realized by Niels Bohr.

He then realized that electromagnetic (light energy) wave must be produced out of something. In this case it is the kinetic energy of the electron!

So energy of motion is getting converted into visible light energy. And if this is the case a time will come when the electron will have no kinetic energy and it will eventually spiral into the nucleus. This will destroy the entire atom!

The orbiting electron will emit energy and thus, lose its kinetic energy. It will slow down and collide with the nucleus.

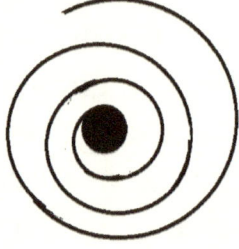

So bad! Atoms would not even exist!

Niels Bohr had to solve this problem (because atoms exist in the real world) and he had to develop an entirely new version of Rutherford's model. He made key changes and his model of atom was accepted...

...for some time!

Anyway, let us talk about electromagnetic waves. These are electric and magnetic fields continuously interacting with each other, thus producing an energetic wave. Electromagnetic waves do not require any medium to travel unlike sound waves or water waves that do require medium.

There are so many different kinds of electromagnetic waves that we use in our daily lives. We listen to the radio, which is actually a group of unique electromagnetic waves. Then there are microwaves that are used in microwave ovens used to heat/cook food.

And don't forget about the colors that you see. They are also electromagnetic waves. But how do we differ between these waves?

There are two properties – wavelength and frequency – associated with any wave. Energy of an electromagnetic wave is directly related to frequency.

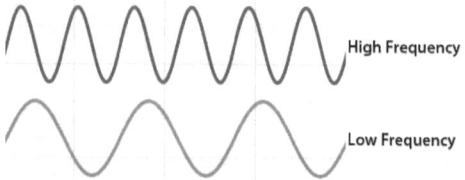

X-rays are high frequency waves while radio and micro waves are low frequency waves. Low frequency waves are less dangerous because they have less energy. On the other hand, high frequency waves like x-rays and gamma rays are dangerous!

Beware!

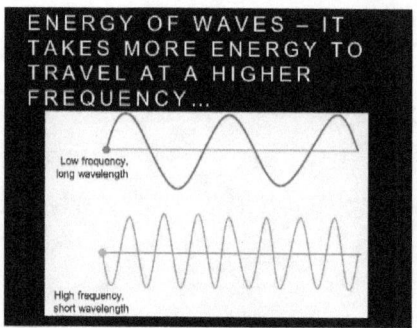

More wavelength suggests a low frequency wave. And a high frequency wave has less wavelength as shown in the image above.

Now that we know about wavelength, frequency and energy, we can understand different types of colors.
The white sunlight that we see is actually made up of a mix of many colors. We have found that they are mainly 7 colors – VIBGYOR – violet, indigo, blue, green, yellow, orange and red…

Now when you rotate this disk about the pencil (axis) they will all mix together (blend together) to produce white color!
So sunlight is actually white light. And we see rainbows because droplets of water divide the white sunlight into different colors!

An interesting question: *what makes up light?* Newton suggested a long time ago that light was made of particles.

But Newton was then challenged by Huygens who said that light was nothing but a wave carrying energy. Today we know that light can be thought of as both, but that is another complicated issue...

Into the Universe

We have discussed the tiny world of atoms. We also looked at the macroscopic world when we learned about motion of objects in our everyday lives. We studied about the two fundamental forces of nature – gravity and electromagnetism. Now we have to go a step further – into the universe!

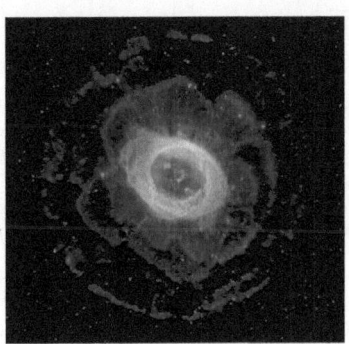

What is the universe made of? Empty space, stars and galaxies! Actually galaxies are groups of millions of stars! Our understanding of the universe began when we realized that Copernicus and Galileo were right. They said that sun was at the center and planets like earth moved around it. This is called our *solar system*. Solar system is really big!

Then we have galaxies. Sun is a star that is revolving around the center of the Milky Way galaxy. There may be millions of galaxies in the universe, maybe more...

There was a time when we thought that ours was the only galaxy in the universe and all the other shining things that we saw were other stars.

But with powerful telescopes, we realized that there were other galaxies too! It was Edwin Hubble who first stated that Milky Way was not the only galaxy in the universe.

We then asked ourselves,

*"If there are other galaxies,
then there will be other stars
too. And if there are other
stars like our sun, there will be
planetary systems too. And if
there are planetary systems,
there will be planets like earth
too…"*

We got excited!

This led to a series of space tours and the age of space exploration began. We first started with the moon. And then we sent probes to Mars, Venus and then Jupiter and Saturn.

We did not find life anywhere else in our solar system. So we started finding life somewhere else. But we have not found it yet, after all these years. Scientists hope that in this vast universe there has to be enough appropriate conditions for life to exist on other planets…

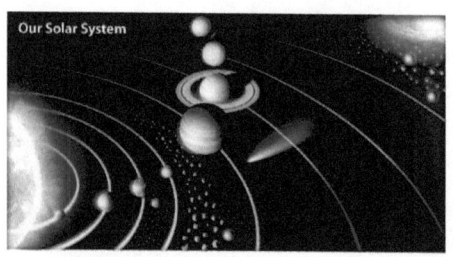

Now let us talk about stars. How do stars make light? Why do they shine? Actually stars are using the principle of conversion of energy (conservation). They are like factories where light is made. Every factory needs some investment to start making goods.

And so, stars use *hydrogen*. Hydrogen has mass. Stars burn hydrogen (they burn mass) and convert mass-energy into light energy. Actually this so-called mass energy is directly derived from Einstein's *energy-mass equivalence*.

So stars are burning themselves up and producing light! This means that stars can also run out of fuel – if not today, then maybe someday! If this ever happens, a star will meets its end.

So stars also have a life-cycle! They grow old and die...

Stars also come in different colors. We can look at stars and tell how bright they are by identifying their color! Astronomers use this idea to measure distances to stars. If a star is bright, it must be closer than another star which appears dim. Simple application!

Another important question regarding our universe is this,

"When did the universe come into existence?"

This is a very difficult question. We humans have always been curious. We have learned so many things from nature. But now we are questioning our own existence and our own future.

Will we ever know the correct absolute answer to how the universe came into being? We do not know that yet...

But think about this,

If we come to know the answer someday by chance, won't we be tempted to create our own miniature universe by following exactly the same process?

That seems interesting and bizarre at the same time...

About the author

Vedang Sati *(Illustrator, Physics Enthusiast)* is the founder of WonderPHY6, an online science community primarily interested in the promotion of physics to younger audiences. He was awarded Student of the Year award (2013) by The Times of India Newspaper in Education (NiE) for excellent academic and non-academic performance.

Vedang started working as a science promoter when he was only 16. And to interact with more number of people, he set up an online medium called WonderPHY6 in 2014. As of today, WonderPHY6 has a weekly reach of over 80,000 unique people (source - Facebook Analytics). Currently, he is an Engineering undergraduate student studying the rich fundamentals of electronics.

Curious?
Visit: **www.facebook.com/wonderphy6**